Solving problems using electronic circuits

AND using diodes and pull-up resistor 3

AND using transistors 4

Circuit for determining the static current gain coefficient of a transistor 6

Circuit with stabilization in the emitter circuit 7

Diode limiter 8

Diode limiter with Zener diode 13

Energy and power — why resistors burn 15

Full-wave rectifier 16

Half-wave rectifier and voltage indicator 17

Half-wave rectifier 19

How to get high amplifier input impedance? 20

How to increase the current gain of a transistor? 22

How to increase the reverse voltage of a diode? 22

How to increase the output current of a diode? 23

How to lower AC voltage? 24

How to match the load resistance to the source resistance? 25

How to measure current and voltage? 26

Increasing and decreasing the value of electronic components 28

Meter overload protection 29

NOT using transistors 30

Ohm's and Kirchhoff's voltage Law 31

OR using diodes and pull-down resistor 34

OR using transistors 35

Phase-splitter 37

Rectifier in a bridge configuration 40

Series and parallel connection of capacitors 41

When will the power transferred from the source to the load be greatest? 42

Voltage divider and loaded voltage divider 42

Additions 43

Solving problems using electronic circuits

Addition A. Metric prefixes 43

Addition B. Decibels 45

Addiction C. How to read oscilloscope parameters? 46

Addition D. Effective value of alternating current 47

Solving problems using electronic circuits

AND using diodes and pull-up resistor

Below input In1=H, In2=H. Output Out=H.

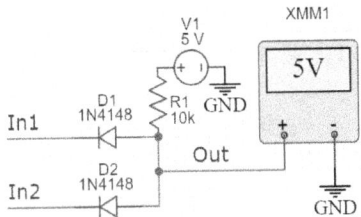

Below input In1=L, In2=H. Output Out=L.

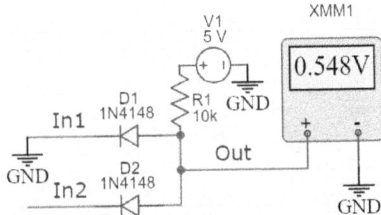

Below input In1=H, In2=L. Output Out=L.

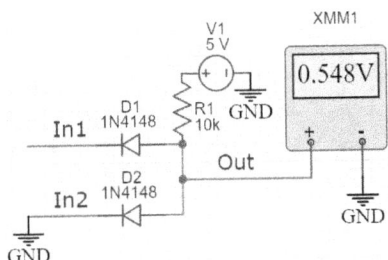

Below input In1=L, In2=L. Output Out=L.

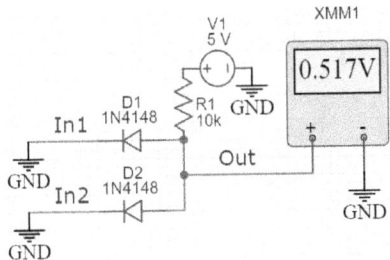

This circuit is logical AND. Output=H only if Input1=H and Input2=H.

Solving problems using electronic circuits

Practical note. The above circuit is perfect for explaining the principle of operation of an AND gate built using diodes. Leaving the end of the element unconnected causes it to act as an antenna and voltages may be induced on it by external magnetic fields. In practice, it is better to support it with a resistor (R5 and R6) connected to the supply voltage (V2), and force the low state by pressing a keys (K1 and K2) connected to ground (picture below).

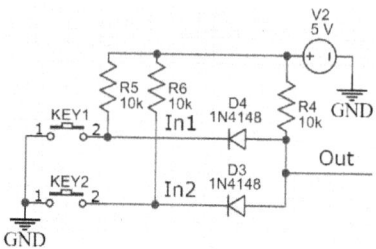

Practical note. Need an AND gate with more inputs? Nothing could be simpler than duplicating the existing pattern the appropriate number of times (picture below).

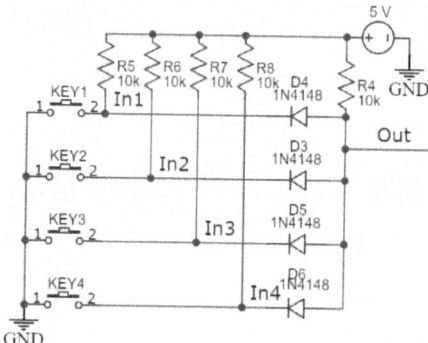

AND using transistors

Below input In1=H, In2=H. Output Out=H.

Solving problems using electronic circuits

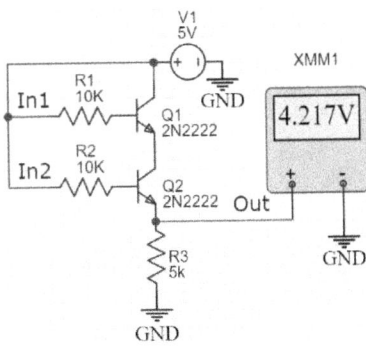

Below input In1=L, In2=H. Output Out=L.

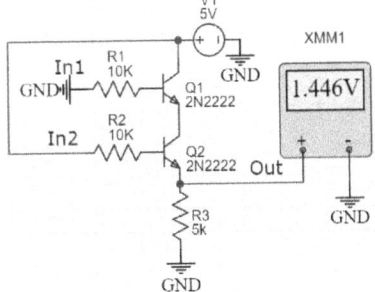

Below input In1=H, In2=L. Output Out=L.

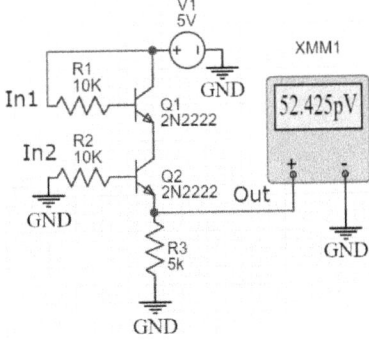

Below input In1=L, In2=L. Output Out=L.

Solving problems using electronic circuits

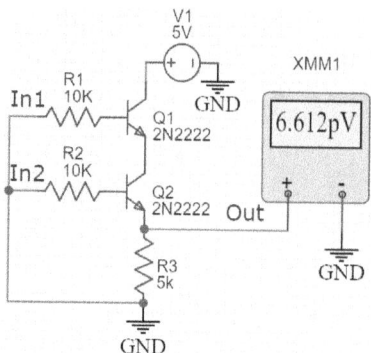

This circuit is logical AND. Output=H only if Input1=H and Input2=H.

Circuit for determining the static current gain coefficient of a transistor

The collector current can be controlled by the base current. A small base current has a small effect on the depletion layer, and the collector current remains small.

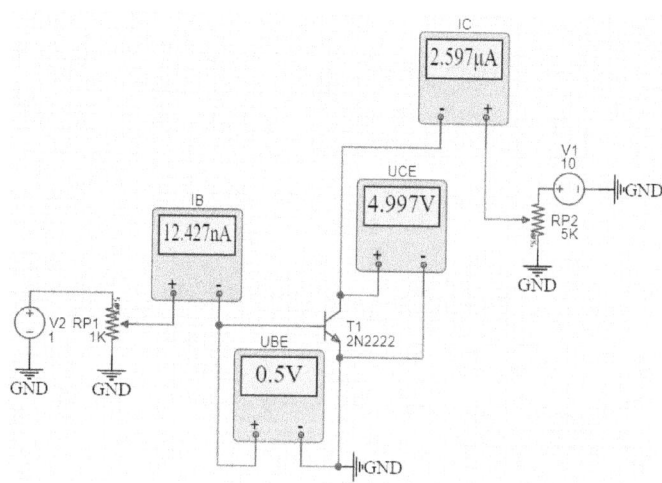

$$\beta = \frac{IC}{IB}$$

There is

Solving problems using electronic circuits

$$\beta = \frac{2{,}579\,\mu A}{12{,}427\,nA} = 208$$

As the value of the base current changes using potentiometer P1 the collector current changes proportionally. It becomes significantly larger.

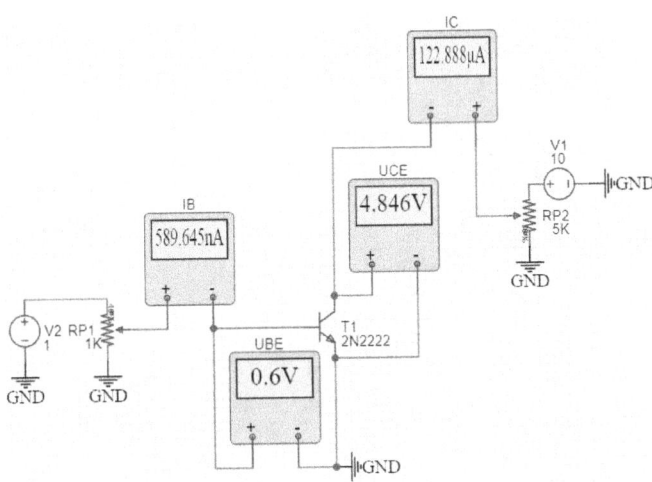

Circuit with stabilization in the emitter circuit

In figure below, a bias circuit with stabilization in the form of a resistor connected in the emitter circuit is shown. In a simple circuit without this resistor, an increase in the collector current I_{co} causes an increase in the collector current and increases the voltage drop across the resistor in the base circuit, which causes further positive biasing of the junction and a further increase in emitter and collector currents. Adding a resistor in the emitter circuit counteracts the increase in currents because any momentary increase in current causes an increase in the voltage drop across this resistor, and hence an increase in the bias voltage in the non-conducting direction; this in turn causes a decrease in the current increase, thus stabilizing its level at a certain, almost constant level.

Solving problems using electronic circuits

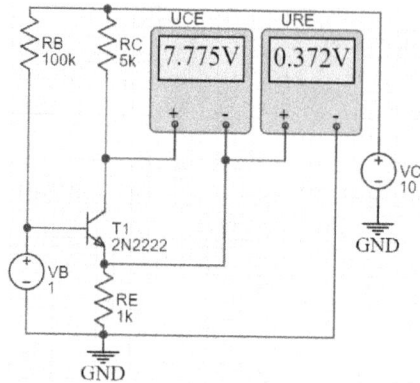

Increase VB and decrease UCE:

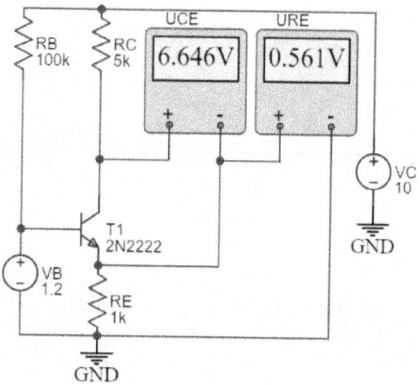

Diode limiter

Circuit 1st

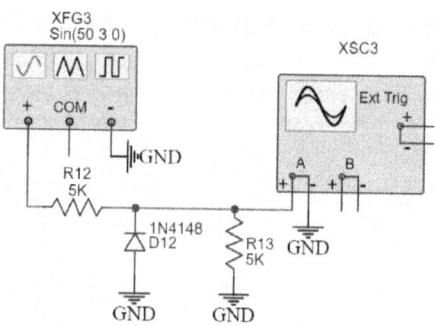

Oscillogram 1st

Solving problems using electronic circuits

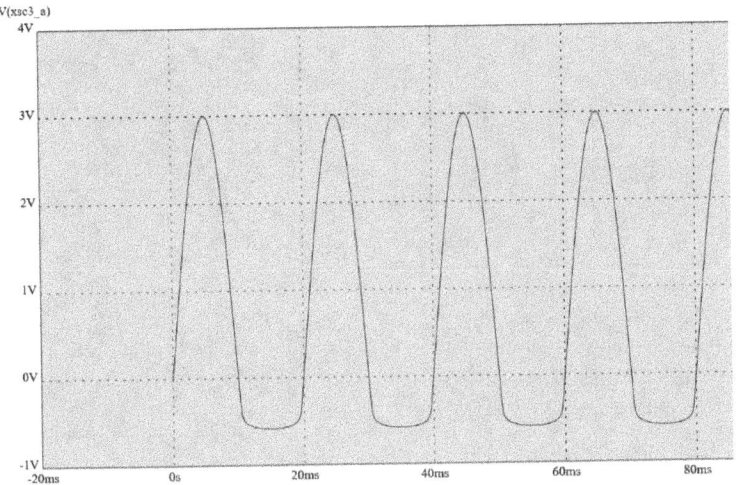

Circuit 2nd

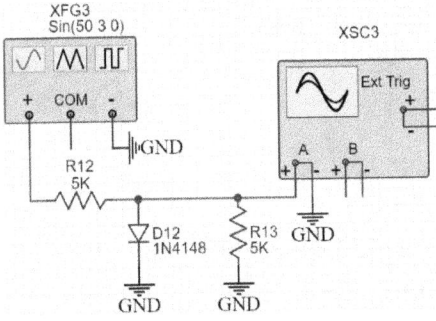

Oscillogram 2nd

Solving problems using electronic circuits

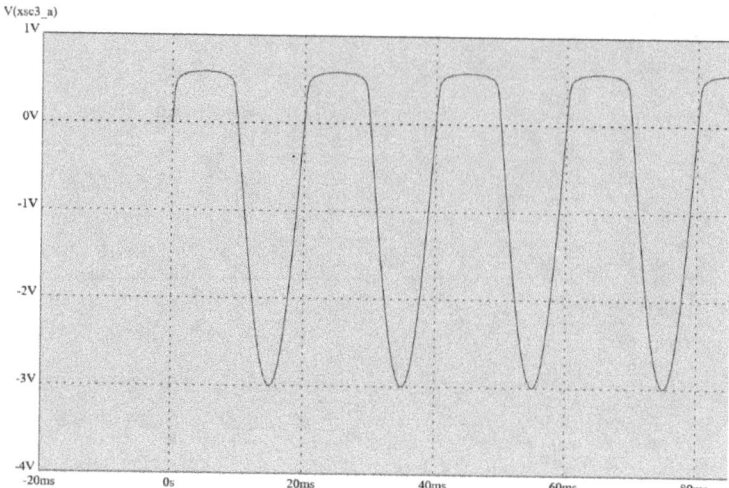

Circuit 3rd

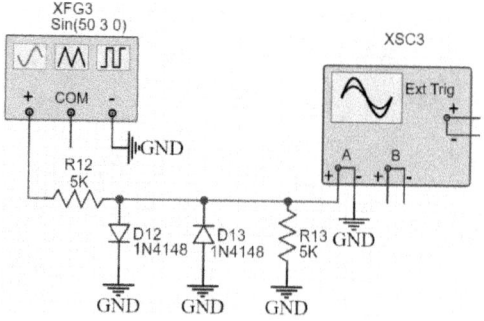

Oscillogram 3rd

Solving problems using electronic circuits

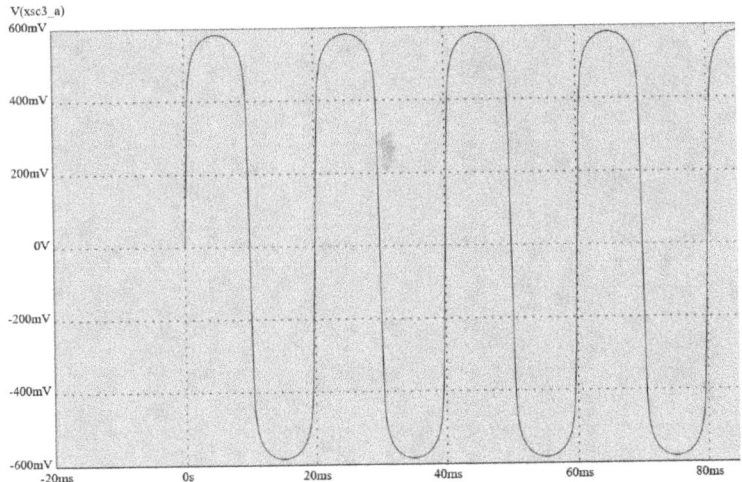

Circuit 4th

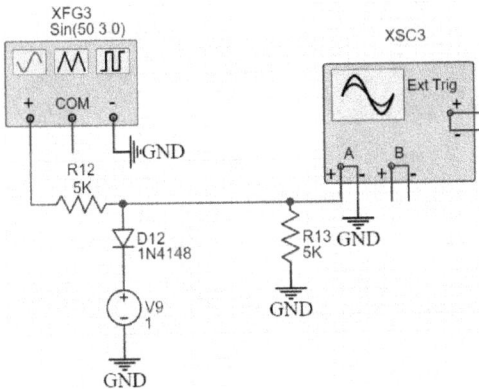

Oscillogram 4th

Solving problems using electronic circuits

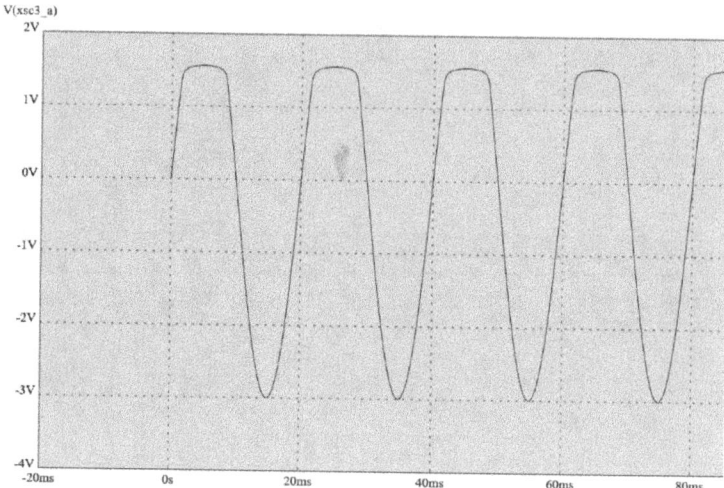

Circuit 5th

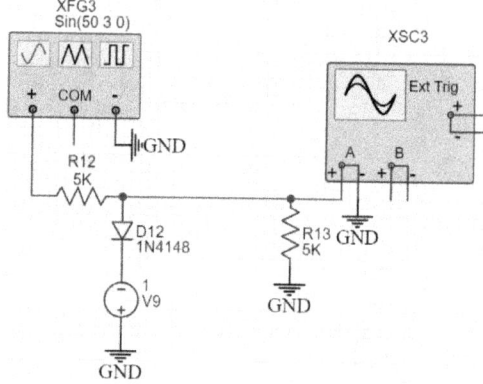

Oscillogram 5th

Solving problems using electronic circuits

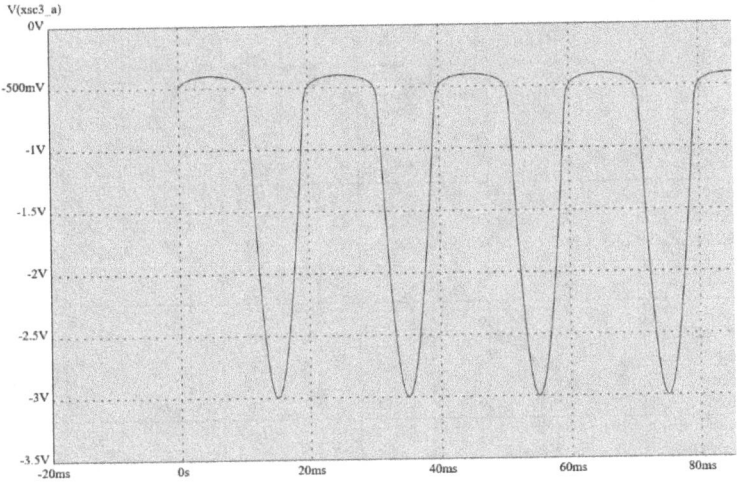

Diode limiter with Zener diode

Circuit 1st

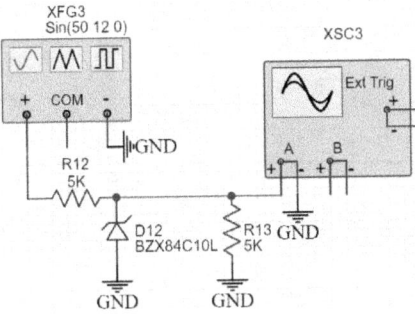

Datasheet od diode (Source: https://pdf1.alldatasheet.com/datasheet-pdf/view/71139/LRC/BZX84C10L.html)

Solving problems using electronic circuits

ELECTRICAL CHARACTERISTICS
(Pinout: 1-Anode, 2-NC, 3-Cathode) (V$_F$ =0.9V Max @ I$_F$ =10 mA for all types)

Type Number	Marking	Zener Voltage Vz1(Volts) @IZT1=5mA (Note 1)			Max Zener Imped-ance Z$_{zT}$ (Ohms) @I$_{zT}$=5mA	Max Reverse Leakage Current		Zener Voltage V$_{zz}$(Volts) @I$_{zT}$=1mA (Note1)		Max Zener Impedance Z$_{zT}$ (Ohms) @I$_{zT}$=5mA	Zener Voltage Vz1(Volts) @IZT1=5mA (Note 1)		Max Zener Imped-ance Z$_{zT}$ (Ohms) @I$_{zT}$=20mA	d$_{vz}$/dt (mv/k)		CpF Max @VR=0 @IZT1=5mA f=1MHz
		Nom	Min	Max		I$_R$ @ uA	V$_R$ Volts	Min	Max		Min	Max		Min	Max	
BZX84C2V4LT1	Z11	2.4	2.2	2.6	100	50	1	1.7	2.1	600	2.6	3.2	50	-3.5	0	450
BZX84C2V7LT1	Z12	2.7	2.5	2.9	100	20	1	1.9	2.4	600	3	3.6	50	-3.5	0	450
BZX84C3V0LT1	Z13	3	2.8	3.2	95	10	1	2.1	2.7	600	3.3	3.9	50	-3.5	0	450
BZX84C3V3LT1	Z14	3.3	3.1	3.5	95	5	1	2.3	2.9	600	3.6	4.2	40	-3.5	0	450
BZX84C3V6LT1	Z15	3.6	3.4	3.8	90	5	1	2.7	3.3	600	3.9	4.5	40	-3.5	0	450
BZX84C3V9LT1	Z16	3.9	3.7	4.1	90	3	1	2.9	3.5	600	4.1	4.7	30	-3.5	-2.5	450
BZX84C4V3LT1	W9	4.3	4	4.6	90	3	1	3.3	4	600	4.4	5.1	30	-3.5	0	450
BZX84C4V7LT1	Z1	4.7	4.4	5	80	3	2	3.7	4.7	500	4.5	5.4	15	-3.5	0.2	260
BZX84C5V1LT1	Z2	5.1	4.8	5.4	60	2	2	4.2	5.3	480	5	5.9	15	-2.7	1.2	225
BZX84C5V6LT1	Z3	5.6	5.2	6	40	1	2	4.8	6	400	5.2	6.3	10	-2.0	2.5	200
BZX84C6V2LT1	Z4	6.2	5.8	6.6	10	3	4	5.6	6.6	150	5.8	6.8	6	0.4	3.7	185
BZX84C6V8LT1	Z5	6.8	6.4	7.2	15	2	4	6.3	7.2	80	6.4	7.4	6	1.2	4.5	155
BZX84C7V5LT1	Z6	7.5	7	7.9	15	1	5	6.9	7.9	80	7	8	6	2.5	5.3	140
BZX84C8V2LT1	Z7	8.2	7.7	8.7	15	0.7	5	7.6	8.7	80	7.7	8.8	6	3.2	6.2	135
BZX84C9V1LT1	Z8	9.1	8.5	9.6	15	0.5	6	8.4	9.6	100	8.5	9.7	8	3.8	7.0	130
BZX84C10LT1	Z9	10	9.4	10.6	20	0.2	7	9.3	10.6	150	9.4	10.7	10	4.5	8.0	130

Oscillogram 1st

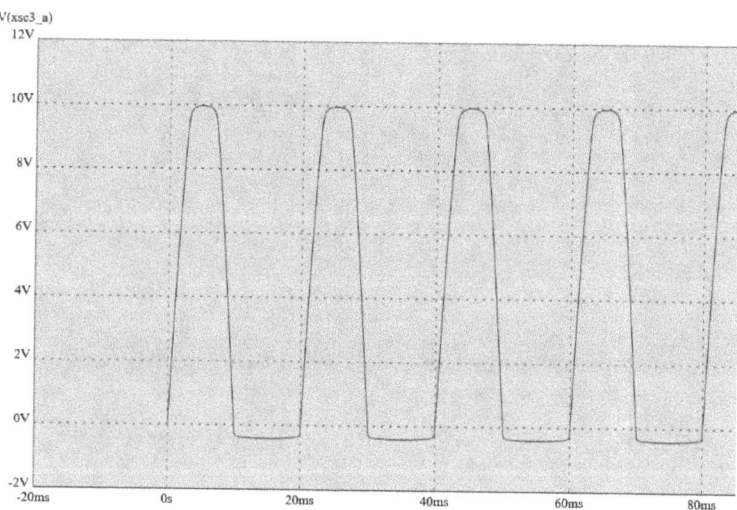

Solving problems using electronic circuits

Energy and power — why resistors burn

In electrical systems, energy can be transferred by electrons, if there is an EMF and the circuit is closed. It can be calculated using the formula:

$$E_n = U * Q$$

where:

E_n — energy in joules (watt-seconds),

U — expressed in volts, and

Q — in coulombs.

The „rate" of energy change used in electrical systems is called power. The concept of rate includes time, e.g., km/h or miles/h. Therefore, power is energy per unit of time, which we express with the formula:

$$P = \frac{U * Q}{t}$$

where:

P — power in watts (W), and

t — time in seconds.

If the rate of current flow is equal to coulombs per second (i.e., A = C/s) and if P=U(Q/t), then, assuming I in amperes, where I = Q/t , we get P = UI. This is the basic formula for defining power. It can be transformed into the following forms:

$$P = U * I$$

$$U = \frac{P}{I}$$

$$I = \frac{P}{U}$$

Therefore, the power that will be dissipated in the resistor R2 (below)

Solving problems using electronic circuits

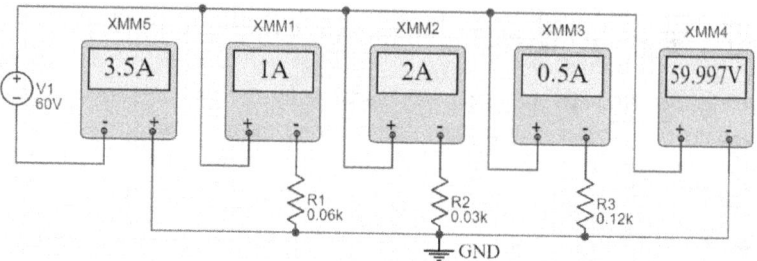

equals

$$P = 60V * 2A = 120W$$

Wow! That's a lot. Most manufacturers specify the power rating at 70°C and free airflow conditions. At temperatures above 70°C, the resistor is derated (picture below – source https://eu.mouser.com/c/ds/passive-components/resistors/).

MODEL	RESISTANCE RANGE Ω	RATED POWER $P_{70°C}$ W
RCMM02	1 to 332K	0.25
	1 to 332K	0.50
RCMM05	1 to 1M	0.50
RCMM1	1 to 2.26M	1.0

STANDARD ELECTRICAL SPECIFICATIONS

Practical note. Derating is a design technique where components are operated at less than their rated maximum parameters. This reduces the degradation rate and increases the component's life expectancy and reliability. The recommended value is 80% maximum parameters for fixed resistors and 75% for variable resistors.

Full-wave rectifier

A full-wave or two-cycle rectifier is called such a rectifier in which, after the rectification process, only those parts of the waveform that are of one sign remain, while the parts of the opposite sign are eliminated. The basic circuit of a full-wave rectifier controlled by a sinusoidal signal with a transformer is shown below

Solving problems using electronic circuits

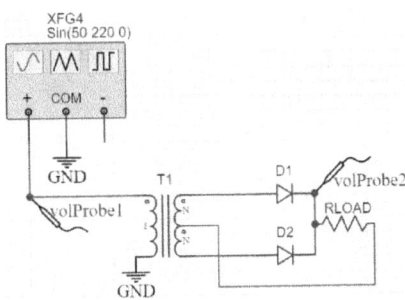

During the periods when a positive voltage appears on the anode of diode D1, the situation is such that the potential at the secondary winding of the transformer is positive relative to the central tap. During the positive half-cycle, we have a situation where diode D1 conducts, while diode D2 does not conduct. During the negative half-cycle, the potential at the secondary winding reverses, and diode D2 conducts, while diode D1 does not conduct. As a result, we get a pulsating DC voltage at the output, which is greater than the voltage obtained at the output of a half-wave rectifier.

The average value of the output voltage from the full-wave rectifier — as is easy to notice — is greater than the voltage obtained at the output of a half-wave rectifier.

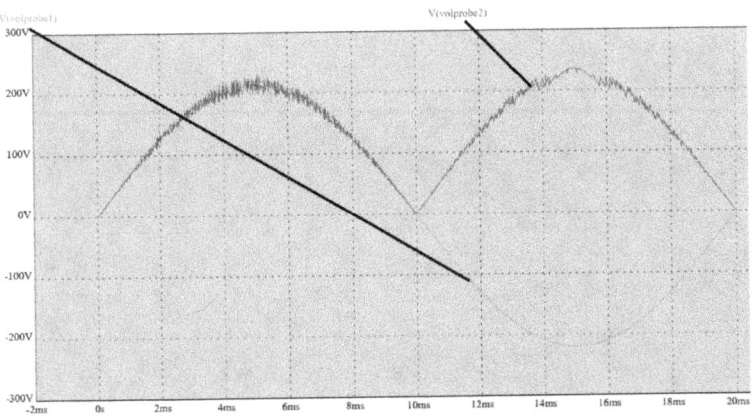

Half-wave rectifier and voltage indicator

Half-wave rectifier in the measuring circuit

Solving problems using electronic circuits

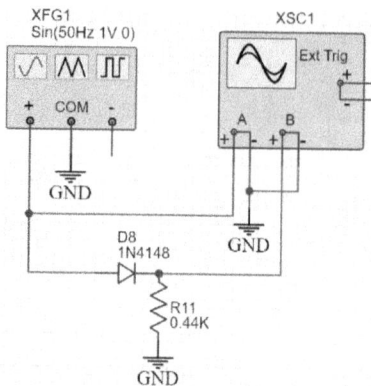

Oscillograms at the input and output of a half-wave rectifier.

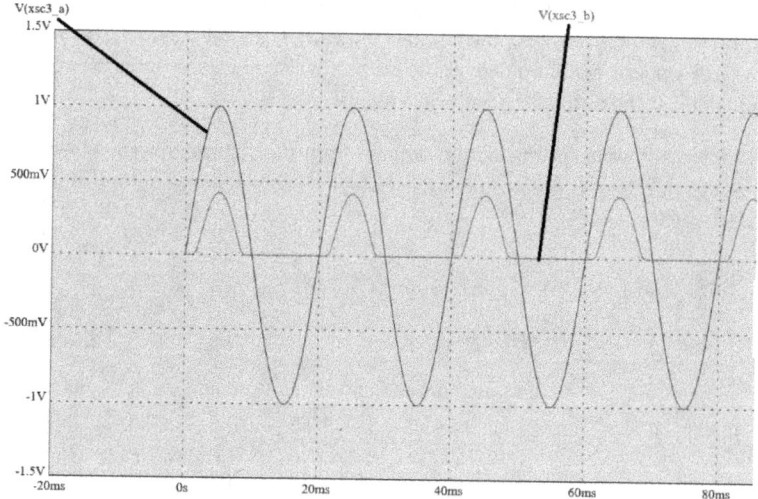

Below is voltage indicator.

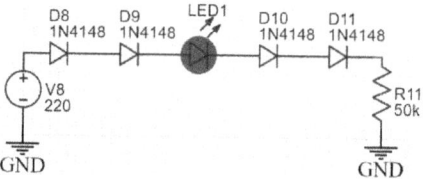

Each 1N4148 diode can withstand a reverse voltage of up to 75V (the datasheet from https://pdf1.alldatasheet.com/datasheet-pdf/view/15021/PHILIPS/1N4148.html below). Since the mains voltage is once +220V and once -220V, four diodes must be used (3 * 75V = 225V and a 25%

spare in the form of the fourth diode) so that the permissible voltage is not exceeded in the cut-off state. If we ignore the voltage drop on the conducting diodes (4 * 0.7V = 2.8V + about 1 V LED), the current flowing through the diodes will be 220V: 50k = 4.4mA

High-speed diodes   1N4148; 1N4448

FEATURES
- Hermetically sealed leaded glass SOD27 (DO-35) package
- High switching speed: max. 4 ns
- General application
- Continuous reverse voltage: max. 75 V
- Repetitive peak reverse voltage: max. 75 V
- Repetitive peak forward current: max. 450 mA.

APPLICATIONS
- High-speed switching.

DESCRIPTION
The 1N4148 and 1N4448 are high-speed switching diodes fabricated in planar technology, and encapsulated in hermetically sealed leaded glass SOD27 (DO-35) packages.

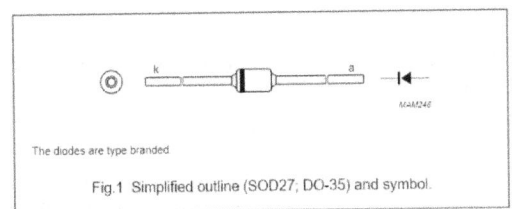

The diodes are type branded

Fig.1 Simplified outline (SOD27; DO-35) and symbol.

Diodes conduct only during their positive half cycle. In the negative half cycle, the LED does not light. All in all, you can see a flicker indicating the presence of voltage.

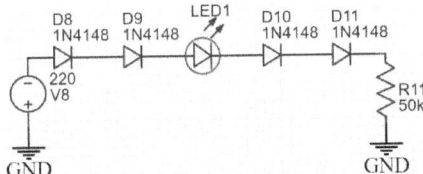

Half-wave rectifier

A half-wave or half-cycle rectifier is called such a rectifier in which, after the rectification process, only those parts of the waveform that are of one sign remain, and the parts of the opposite sign are eliminated. The circuit of a half-wave rectifier controlled by a sinusoidal signal is shown below.

Solving problems using electronic circuits

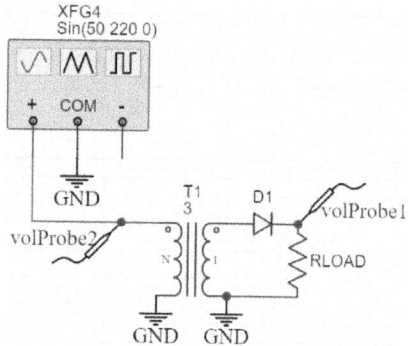

The diode is connected in such a way that it conducts only for the positive half-cycles of the input waveform, as only then is the voltage positive at its anode greater than the potential of the cathode.

In the negative half-cycle, the diode does not conduct and the entire voltage applied to the rectifier appears on the diode as the so-called reverse voltage of the rectifier. When the direction of connection is reversed, the diode will conduct only for the negative half-cycles and will not conduct for the positive half-cycles.

Oscillogram:

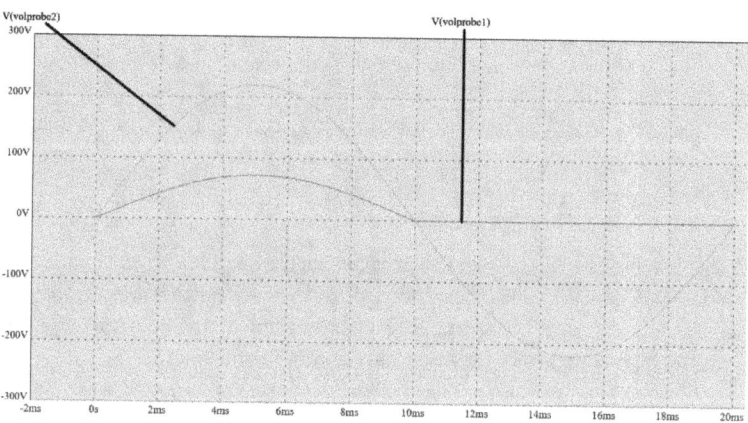

How to get high amplifier input impedance?

In transistor circuits, with the exception of circuits with unipolar transistors characterized by a naturally high input impedance, achieving a significant level of input impedance is much more challenging. The input resistance of an amplifier operating in the common emitter configuration does not exceed

Solving problems using electronic circuits

several tens of kiloohms. Therefore, to achieve high input impedances, special circuits must be used.

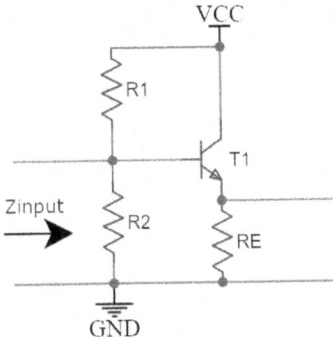

The input impedance of the emitter follower is expressed by a simple formula:

$$Z_{input} \approx \beta * RE$$

which means that it is equal to the resistance in the emitter circuit multiplied by the current gain coefficient (beta) of the transistor.

In the case of transistors operating in the "super-alpha" configuration (below),

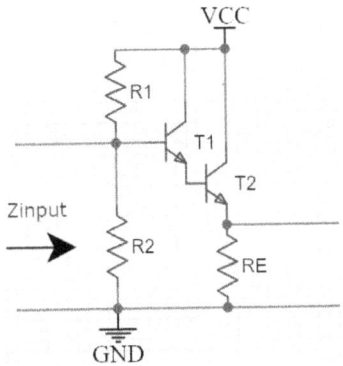

the emitter current of the first transistor controls the base of the second transistor, and so on. The resultant current gain coefficient β is equal to the product of the current gain coefficients β1*β2 of the individual transistors. This can also be achieved using the Darlington circuit. In this configuration, the input impedance of the emitter follower is expressed by the formula:

$$Z_{input} \approx \beta 1 * \beta 2 * RE$$

How to increase the current gain of a transistor?

The signal fed to the input of transistor T2 is again amplified by transistor T3. At the output of the system, we obtain a gain that is the product of the gains of both transistors

$$\beta_{whole} = \beta_{T2} * \beta_{T3}$$

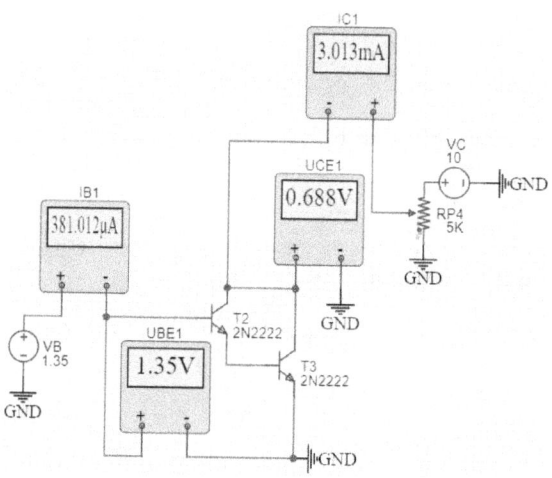

How to increase the reverse voltage of a diode?

Characteristic	Symbol	1N 4001/L	1N 4002/L	1N 4003/L	1N 4004/L	1N 4005/L	1N 4006/L	1N 4007/L	Unit
Peak Repetitive Reverse Voltage Working Peak Reverse Voltage DC Blocking Voltage	V_{RRM} V_{RWM} V_R	50	100	200	400	600	800	1000	V
RMS Reverse Voltage	$V_{R(RMS)}$	35	70	140	280	420	560	700	V

Why increase the reverse voltage of a diode? 230V after rectification can give as much as 324V, and taking into account the permissible fluctuations of the mains voltage "upwards" by 10%, it can be even 357V. So, the reverse voltage on the diodes will reach over 700V! Therefore, for example, 1N4007 diodes are used to rectify the mains voltage, which have a permissible reverse voltage of 1000V. What to do if we do not have such diodes? You can connect diodes with a lower reverse voltage in series. The problem is that although the diodes look the same and are of the same type, they differ slightly in parameters. To protect the weakest diode, a parallel resistor must be connected in parallel to each diode. Its resistance should be lower than the diode leakage for the worst voltage and temperature conditions. For BYP680-300 diodes, e.g. 100 - 330k may be enough. Why? If the leakage is <100uA, then the shunt resistor current is 10 times larger (1mA, because 300V/1mA=300k).

Below is a diode connected in series with shunt resistors

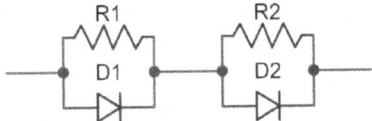

How to increase the output current of a diode?

If there are no higher power rectifier diodes, the diodes can be connected in parallel as he wrote, but:

- the number of diodes should be such that the current through each one does not exceed 0.8 of its allowable current,
- the resistors should have a value such that a voltage drop of at least 50% of the diode's forward voltage is deposited on them.

Of course, all diodes on a common radiator.

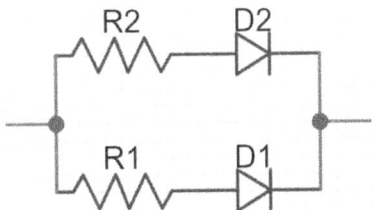

Parallel connection - without additional equalizing resistors - is inappropriate, because any asymmetry in the flow of currents causes this asymmetry to deepen and ultimately - thermal destruction.

This happens because the voltage drop across the conducting diode decreases with the increase in junction temperature (by 2mV per 1°C), then the hotter diode takes on more current than the colder one, so it heats up even more, which "improves" its conduction - until finally the asymmetry is so great that the colder diode practically does not participate in the current flow at all, and the whole thing "crashes" through the hot one, which of course will not withstand a load twice as big.

How to lower AC voltage?

A transformer is an inductive element consisting of at least two windings, designed to transfer energy from the primary winding to the secondary winding. In electronics, the transformer is most commonly used for voltage step-up or step-down, e.g., in rectifier circuits in power supplies, and for matching the load resistance placed on the secondary side to the source resistance on the primary side. Transformers are often used as coupling elements in amplifiers. The ratio of the number of secondary turns n_2 to the number of primary turns n_1 is called the transformer's turn ratio p. In an ideal transformer, i.e., lossless, we have the following relationships:

Turn ratio:

$$p = \frac{n_2}{n_1} = \frac{U_2}{U_1}$$

Circuit:

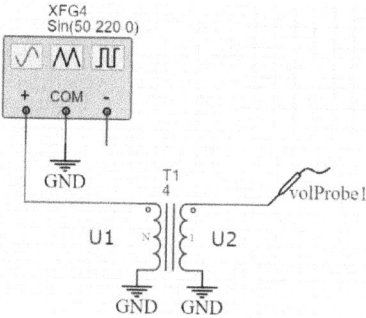

The transformer has a 220V, 50Hz input. With a ratio of 4, the output is still 50Hz, but with an amplitude of 220V:4=55V, so here is output waveform

Solving problems using electronic circuits

oscillogram

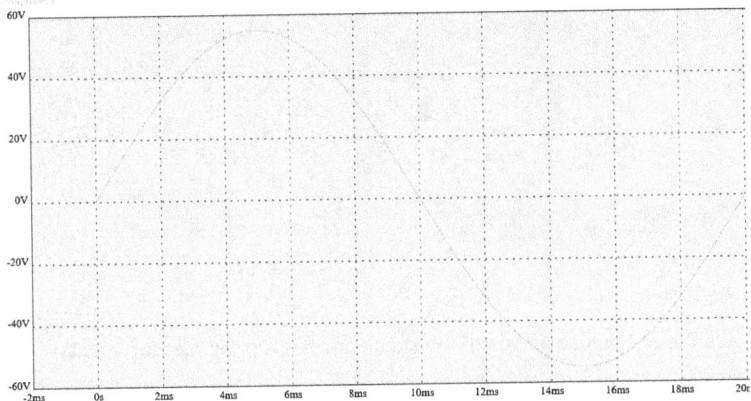

How to match the load resistance to the source resistance?

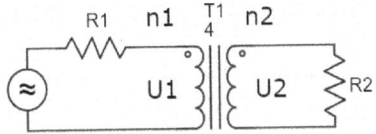

Matching the load resistance R2 to the source resistance R1 occurs when the resistance "seen" from the source side R', or as it is said — transferred to the primary side of the transformer and dependent on the transformer's turn ratio, is equal to the source resistance, i.e., when

$$R' = \frac{R_2}{p^2} = R_1$$

Transferred power:

$$P = \frac{U_2^2}{R_2} = p^2 \frac{U_1^2}{R_2}$$

Solving problems using electronic circuits

How to measure current and voltage?

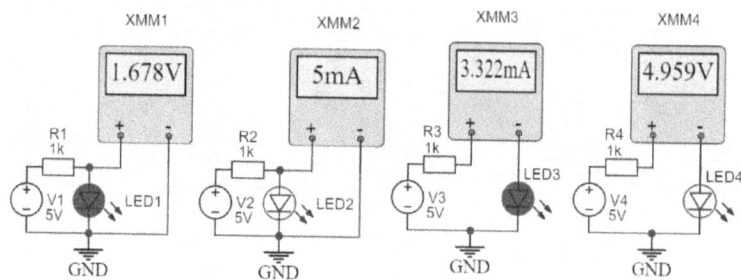

Let's break down each part of the circuit and explain the measurements you're seeing.

Circuit 1 (leftmost)

- **Components**: 5V source (V1), 1kΩ resistor (R1), LED1, and a voltmeter (XMM1).
- **Measurement**: 1.678V across LED1.

Explanation:

The voltmeter XMM1 is measuring the voltage drop across LED1. LEDs typically have a forward voltage drop (the voltage required to light them up), which depends on their color and type. For LED1, it's 1.678V. This means that when a current flows through the LED, it "uses up" 1.678V of the 5V supply, leaving the rest across the resistor.

Circuit 2

- **Components**: 5V source (V2), 1kΩ resistor (R2), LED2, and an ammeter (XMM2).
- **Measurement**: 5mA current through the circuit.

Explanation:

The ammeter XMM2 measures the current flowing through the entire circuit. The resistor R2 limits the current, and with a 5V supply, the current can be calculated using Ohm's law (I = V/R). Since the LED has a voltage drop, the remaining voltage (5V - LED drop) is across R2. If we assume a typical LED drop of around 2V, the current is (5V - 2V) / 1kΩ = 3mA, but here we measure 5mA. Why???

Solving problems using electronic circuits

The ammeter should be connected in series. In this circuit, it is connected in parallel with LED2. The ammeter has a negligible internal resistance so as not to disturb the measurement conditions. Current flows where the resistance is lower. In this case, it flowed through the ammeter and bypassed LED2. The diode does not light. The circuit practically includes: source V2 and resistor R2. The calculations are correct, but the ammeter has been incorrectly connected.

Circuit 3

- **Components**: 5V source (V3), 1kΩ resistor (R3), LED3, and an ammeter (XMM3).
- **Measurement**: 3.322mA current through the circuit.

Explanation:

The XMM3 is connected in series. The ammeter XMM3 measures the current. Here, the current is 3.322mA. Using Ohm's law and the measured current, we can find the voltage drop across R3: V = I * R = 3.322mA * 1kΩ = 3.322V. The remaining voltage (5V - 3.322V = 1.678V) should be across LED3. This suggests that the LED3 has a forward voltage of approximately 1.678V, similar to LED1.

Circuit 4 (rightmost)

- **Components**: 5V source (V4), 1kΩ resistor (R4), LED4, and a voltmeter (XMM4).
- **Measurement**: 4.959V.

Explanation:

The voltmeter XMM4 measures the voltage. This drop is almost the full supply voltage (4.959V out of 5V). This means that the voltage drop across LED4 is very small, around 0.041V (5V - 4.959V). This is unusual for a typical LED, which normally has a higher forward voltage drop.

Why???

The XMM4 voltmeter is connected in series. The current has no other path and must flow through it. The voltmeter has a high internal resistance so as not to change the operating conditions of the element on which the voltage drop is measured. In a series circuit, the same current flows through all elements. The greater the resistance of an element, the greater the voltage drop across it. The voltmeter has such a high internal resistance that almost all the voltage is deposited on it. The current in the system is too low for the LED to glow.

Summary

Solving problems using electronic circuits

Remember, LEDs have characteristic forward voltages, typically around 2V for red, 3V for blue, and so on. The resistor limits the current to protect the LED from burning out. The small differences in measurements can be due to the LEDs' manufacturing variations or slight differences in the circuit setup.

If the measurement results differ significantly from those assumed, it is usually due to a damaged element or a connection error.

Voltmeter - measures the voltage drop between the terminals of the element and between them it should be connected (**parallel** to the element).

Ammeter - measures the current flowing through the element and should be connected **in series** with it.

Increasing and decreasing the value of electronic components

Determination of the resultant resistance RS of resistors R1 and R2 connected in series:

$$RS = R1 + R2$$

Determination of the resultant resistance RP of resistors R1 and R2 connected in parallel:

$$RP = \frac{R1 * R2}{R1 + R2}$$

Determination of the resultant capacitance CS of capacitors C1 and C2 connected in series:

$$CS = \frac{C1 * C2}{C1 + C2}$$

Solving problems using electronic circuits

Determination of the resultant capacitance CP of capacitors C1 and C2 connected in parallel:

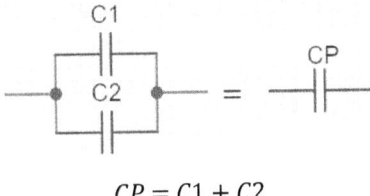

$$CP = C1 + C2$$

Determining the resultant inductance LS of coils L1 and L2 connected in series:

$$LS = L1 + L2$$

Determining the resultant inductance LP of coils L1 and L2 connected in parallel:

$$LP = \frac{L1 * L2}{L1 + L2}$$

Meter overload protection

In the circuit shown below, the Zener diode (10V) is used to protect measuring instruments from overload. As long as the Zener voltage is not exceeded (here 9V), the diode acts as a high-resistance resistor and does not affect the indicator.

Solving problems using electronic circuits

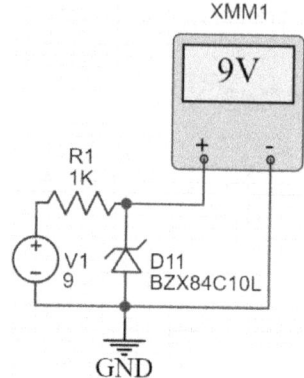

If the input voltage becomes too high (for example 13 V), the diode enters the breakdown region, becomes low-resistant, and shorts the input, thus protecting the instrument from overload.

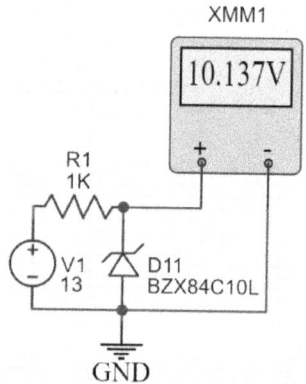

NOT using transistors

Below input is H, output L.

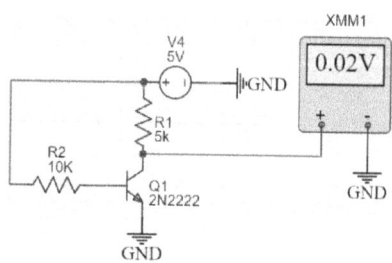

Solving problems using electronic circuits

Below input is L, output H.

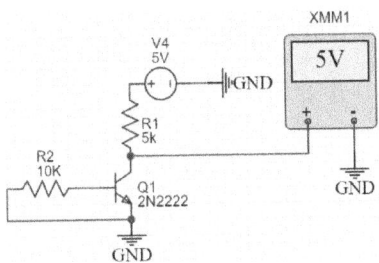

Sure, here's the translation of the provided text from Polish to English:

Ohm's and Kirchhoff's voltage Law

„The value of the current flowing in a circuit is directly proportional to the value of the EMF and inversely proportional to the resistance." This is a simple formulation of Ohm's law. Using mathematical formulas, it can be represented in three forms:

$$I = \frac{V}{R}$$

$$V = I * R$$

$$R = \frac{V}{I}$$

where:

I — current in amperes,

V — EMF in volts, and

R — resistance in ohms.

In the case of a series connection of three resistors, one after the other, figure below).

Solving problems using electronic circuits

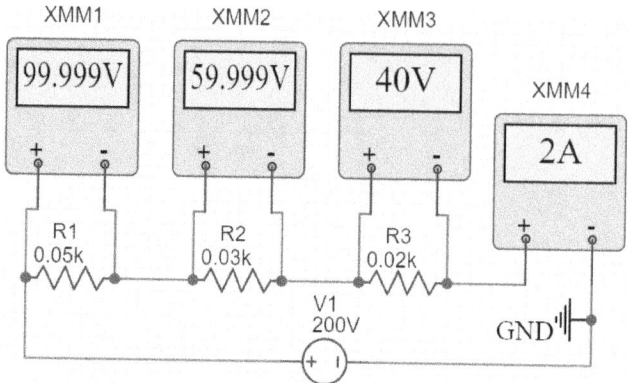

The value of the current flowing through all the resistors and the source V1 is:

$$I = \frac{V}{R}$$

that is

$$I = \frac{200V}{100\Omega}$$

that is

$$I = 2A$$

The voltage drop across resistor R1 caused by the current flow is:

$$V = I * R$$

that is

$$V = 2A * 50\Omega$$

that is

$$V = 100V$$

The voltage drop across the 30 Ω resistor R2 is 60 V, and across R3 — 40 V. The sum of all voltage drops is equal to 200 V, which is the voltage of the source. This is an example of Kirchhoff's voltage law, which states: „**The sum of voltage drops around a circuit is equal to the voltage of the source.**" This law can also be expressed differently: „**The sum of all voltage drops (considering their negative sign) plus the source voltage (considering its positive sign) is equal to zero.**"

Solving problems using electronic circuits

The diagram in figure below shows a load composed of three resistors connected in parallel.

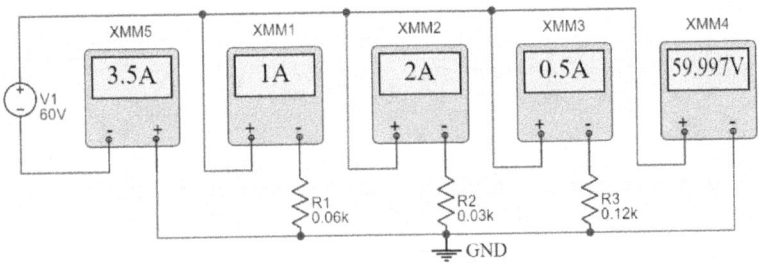

Since R1 is connected to a voltage of 60 V, the current indicated by meter XMM1 is

$$I1 = \frac{V1}{R1}$$

that is

$$I1 = \frac{60V}{60\Omega}$$

that is

$$I1 = 1A$$

The current flowing through R2 is defined as

$$I2 = \frac{V1}{R2}$$

that is

$$I2 = \frac{60V}{30\Omega}$$

that is

$$I1 = 2A$$

The current flowing through R3 is defined as

$$I3 = \frac{V1}{R3}$$

Solving problems using electronic circuits

that is

$$I3 = \frac{60V}{120\Omega}$$

that is

$$I3 = 0.5A$$

Meter XMM4 must be a voltmeter because it is connected in parallel to the circuit. The resistances of ammeters XMM1, XMM2, XMM3, and XMM5 must be very low compared to the resistances of the loads they are connected in series with. In most cases. It is assumed that the resistance of the ammeter is 0 Ω.

The total current flowing in the discussed circuit is the sum of the currents of its three branches and is 1 A + 2 A + 0.5 A, that is 3.5 A, which is indicated by meter XMM5

If the source and wire resistances are negligibly small, the voltmeter should indicate 60 V regardless of whether the three load branches are connected or not. If the internal resistance of the source is not negligible, then a certain voltage drop proportional to the current of all branches will occur across it. The voltage present across the load will be smaller than 60V.

OR using diodes and pull-down resistor

Below input In1=H, In2=H. Output Out=H.

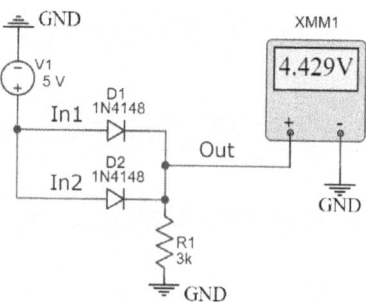

Below input In1=H, In2=L. Output Out=H.

Solving problems using electronic circuits

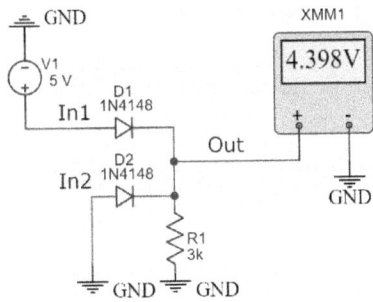

Below input In1=L, In2=H. Output Out=H.

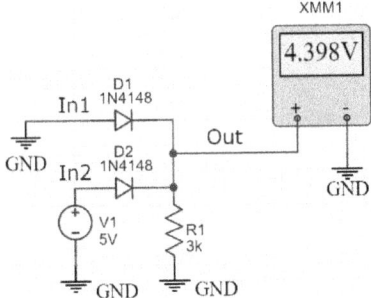

Below input In1=L, In2=L. Output Out=L.

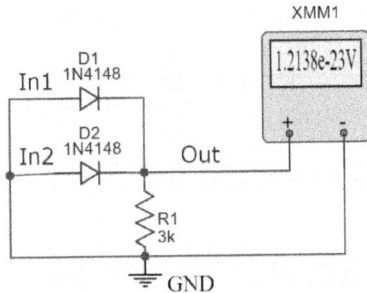

This circuit is logical OR. Output=L only if Input1=L and Input2=L.

OR using transistors

Below input In1=H, In2=H. Output Out=H.

Solving problems using electronic circuits

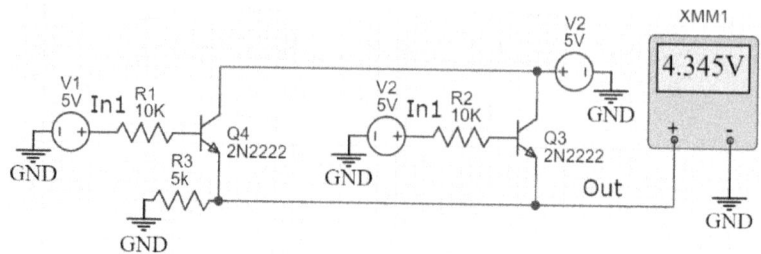

Below input In1=H, In2=L. Output Out=H.

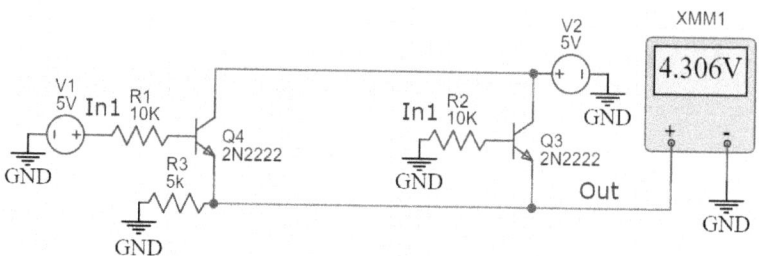

Below input In1=L, In2=H. Output Out=H.

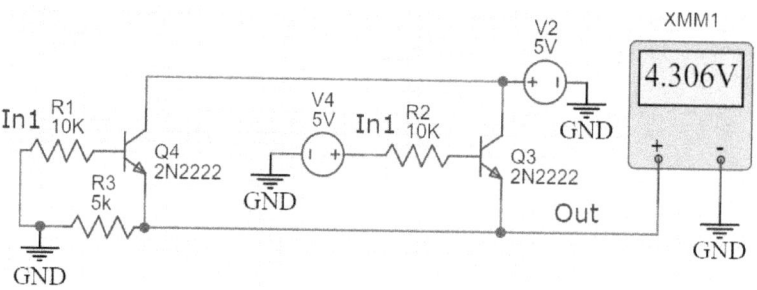

Below input In1=L, In2=L. Output Out=L.

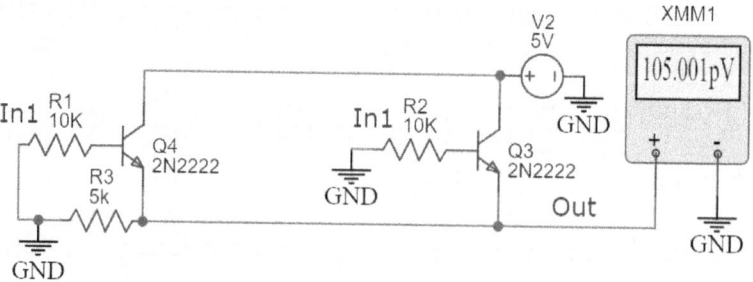

Solving problems using electronic circuits

This circuit is logical OR. Output=L only if Input1=L and Input2=L.

Phase-splitter

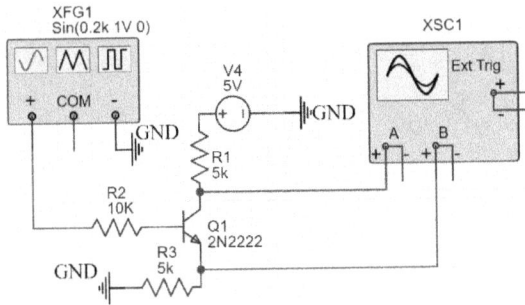

If we combine common collector and common emitter circuits into a single transistor circuit we get the phase splitter.

Now we have two outputs which are 180 degrees out of phase with each other to drive.

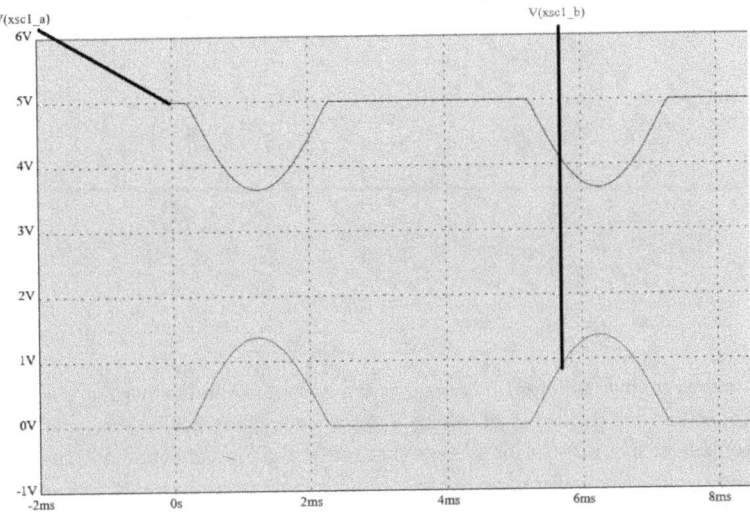

Below is a base polarized circuit that carries the positive and negative halves of the control signal.

Solving problems using electronic circuits

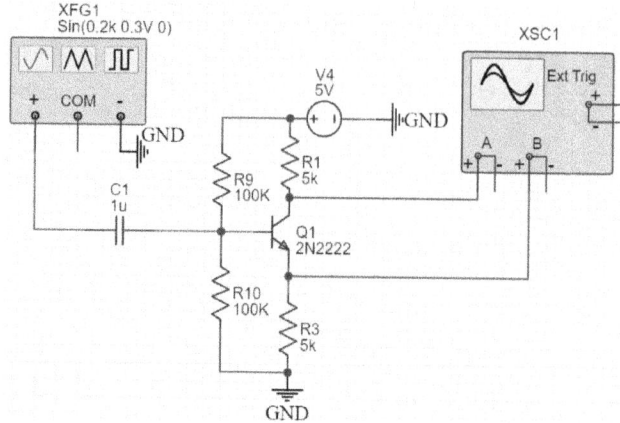

And oscillograms of the signals at the collector and emitter of the transistor.

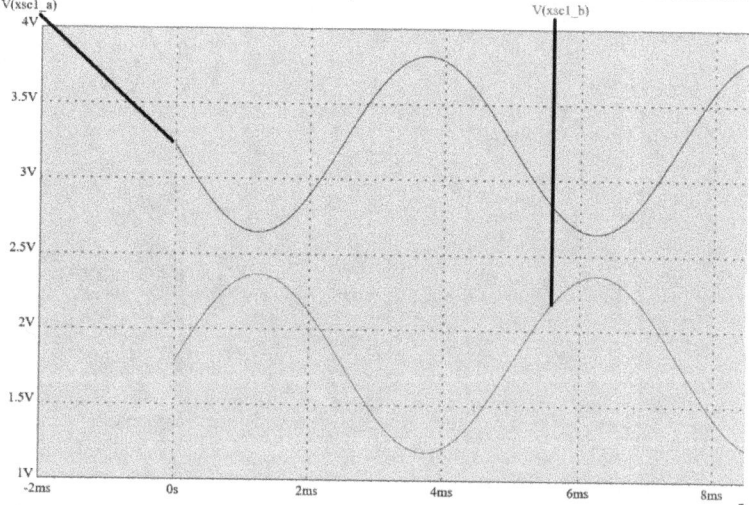

How to remove different DC levels at the outputs? Add the two DC blocking capacitors and resistors tied to +2.5 V as shown in figure below. The capacitors remove or block the DC or average part of the signals and pass the AC part of the signals. The resistors tied to the +2.5 V supply set the new DC or average values for the signals seen at the outputs by charging the capacitors such that the outputs will be centered on +2.5 V.

Practical note. The value of the coupling capacitors set the circuit's low-frequency cutoff point.

Below - circuit

Solving problems using electronic circuits

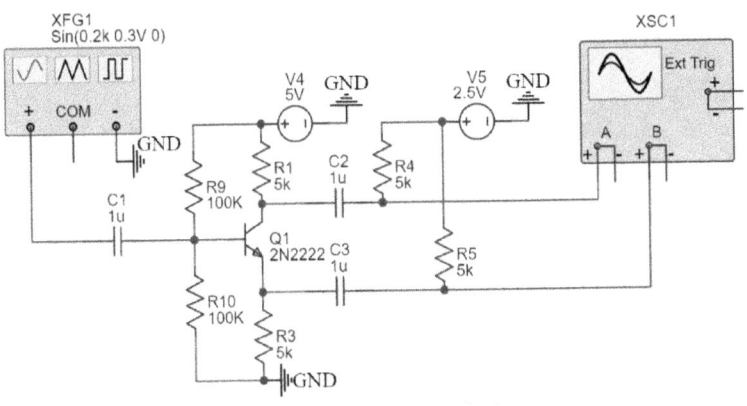

And oscillograms

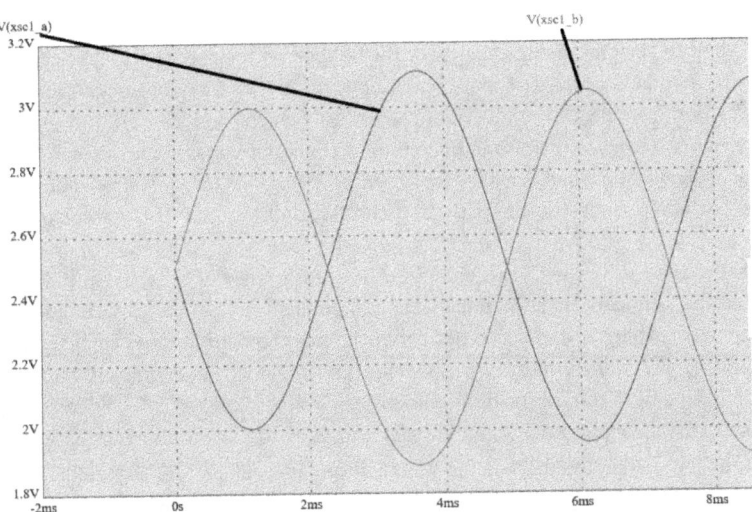

Practical note. Phase splitter driving balanced transmission lines. Balanced transmission lines are used in various communication and audio systems to reduce noise and interference. External electromagnetic interference (EMI) and noise tend to affect both conductors equally (common-mode noise). At the receiving end, a differential amplifier or receiver subtracts one signal from the other. Since the noise is common to both, it cancels out during this subtraction process, leaving the desired signal intact.

Solving problems using electronic circuits

Rectifier in a bridge configuration

A rectifier in a bridge configuration, also known as a Graetz circuit, is a full-wave rectifier with four diodes connected in the manner shown in figure below.

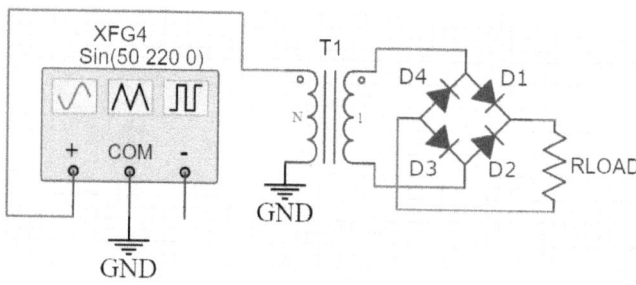

When the instantaneous polarity of the voltage on the secondary winding is as indicated in the figure, diode D1 conducts, and diode D2 does not conduct. At this time, diode D3 conducts in the other branch, and diode D4 does not conduct. In this way, for one half-cycle of the input voltage, the current flows as follows: the end of the transformer being at the negative potential ("bottom"), diode D3, the RLOAD, diode D1, and the end of the transformer being at the positive potential ("top"). At this time, the rectifier is working as a half-wave rectifier with two diodes D3 and D1 connected in series. In the next half-cycle, the polarity of the input voltage changes, and diodes D2 and D4 conduct, while D1 and D3 do not conduct. Now the current flows from the top end of the transformer through diode D4, the load, and diode D2 to the bottom end of the transformer. The system also works as a half-wave rectifier, and the current flowing through the load is in the same direction as in the previous half-cycle. Thus, the current flows through the load in both half-cycles, and in total, the bridge rectifier works as a full-wave rectifier.

Practical note. What rectifier circuit is the most advantageous? The simplest and cheapest is a single-wave rectifier. Its disadvantages are a low output voltage, high ripple factor and low transformer utilization.

The most complicated and expensive is a bridge rectifier. Its advantages are: high output voltage, low ripple factor, good transformer utilization.

A two-diode half-wave rectifier is an intermediate solution in terms of both complexity and parameters.

Solving problems using electronic circuits

Why is the bridge rectifier the most popular? The effort and cost of construction are incurred once. For the entire period of operation of the device, you can enjoy high efficiency of the power supply.

Series and parallel connection of capacitors

The combined capacitance of two capacitors connected in parallel with capacitances of 3 µF each is 6 µF, resulting from the occurrence of twice the larger surface area of the plates. We calculate it using the formula:

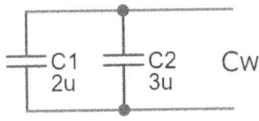

$$Cw = C1 + C2$$

Two capacitors connected in series with a capacitance of 3 µF each have twice the thickness of the dielectric, thus half the capacitance, which is 1.5 µF. We express this using the formula:

$$Cw = \frac{1}{\frac{1}{C1} + \frac{1}{C2}}$$

or

$$Cw = \frac{C1 * C2}{C1 + C2}$$

Solving problems using electronic circuits

When will the power transferred from the source to the load be greatest?

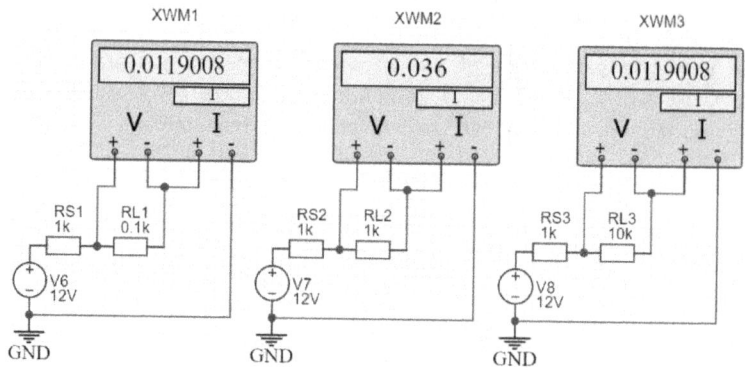

Each system contains: a 12 V source (V), source resistance RS (1 kΩ), load resistance RL (0.1 kΩ, 1 kΩ, 10 kΩ) and a wattmeter (XWM).

If a source with a certain internal resistance (RS) is loaded with an external resistance (RL), it turns out that the power transferred from the source to this load will depend on the value of the external resistance. **Maximum power is delivered when the value of the external load resistance is equal to the value of the source resistance (RL=RS)**. This state is called matching the load to the source.

Voltage divider and loaded voltage divider

Voltage divider (below).

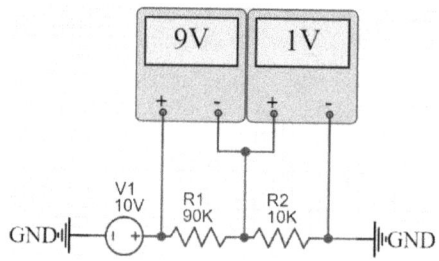

Loaded voltage divider (below).

Solving problems using electronic circuits

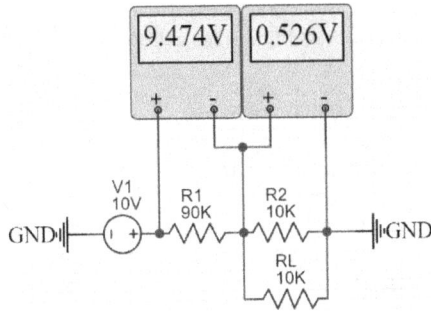

Practical note. A voltage divider (or potentiometer) that is loaded with a parallel resistance RL must have a small resistance value in relation to the load resistance value, otherwise the voltage after connecting the load will be too low.

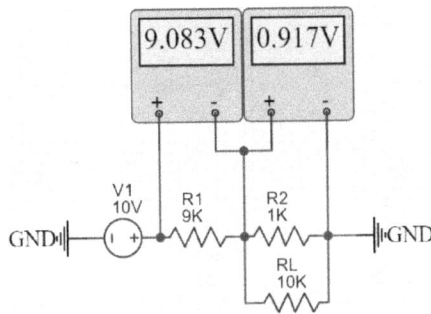

Additions

Addition A. Metric prefixes

This table shows metric prefixes, their symbols, multipliers in both traditional and exponential notation, and their descriptions.

Metric Prefix	Symbol	Multiplier (Traditional Notation)	Exponential	Description
Yotta	Y	1,000,000,000,000,000,000,000,000	10^{24}	Septillion
Zetta	Z	1,000,000,000,000,000,000,000	10^{21}	Sextillion

Solving problems using electronic circuits

Metric Prefix	Symbol	Multiplier (Traditional Notation)	Exponential	Description
Exa	E	1,000,000,000,000,000,000	10^{18}	Quintillion
Peta	P	1,000,000,000,000,000	10^{15}	Quadrillion
Tera	T	1,000,000,000,000	10^{12}	Trillion
Giga	G	1,000,000,000	10^{9}	Billion
Mega	M	1,000,000	10^{6}	Million
kilo	k	1,000	10^{3}	Thousand
hecto	h	100	10^{2}	Hundred
deca	da	10	10^{1}	Ten
base	b	1	10^{0}	One
deci	d	1/10	10^{-1}	Tenth
centi	c	1/100	10^{-2}	Hundredth
milli	m	1/1,000	10^{-3}	Thousandth
micro	µ	1/1,000,000	10^{-6}	Millionth
nano	n	1/1,000,000,000	10^{-9}	Billionth
pico	p	1/1,000,000,000,000	10^{-12}	Trillionth
femto	f	1/1,000,000,000,000,000	10^{-15}	Quadrillionth
atto	a	1/1,000,000,000,000,000,000	10^{-18}	Quintillionth
zepto	z	1/1,000,000,000,000,000,000,000	10^{-21}	Sextillionth
yocto	y	1/1,000,000,000,000,000,000,000,000	10^{-24}	Septillionth

Solving problems using electronic circuits

Addition B. Decibels

An important unit used in electronics is the decibel (dB). It is a measure of the ratio of two powers. We say that the power gain of an amplifier is tenfold if, as a result of applying a 1 W signal to its input, we get a 10 W signal at the output. We can also say that it is 1 bel. 1 bel = 10 decibels, because „deci" means one-tenth. A hundredfold power gain corresponds to achieving ten times tenfold gain, which is logarithmically 20 dB. The numerical representation of the power ratio in decibels is expressed by the formula:

$$dB = 10 \log_{10} \frac{P2}{P1}$$

where

P2 is the output power, and

the base of the logarithm is 10.

In systems where power attenuation occurs, for example, if 1 W of input power results in 0.1 W of output power, the power ratio remains 10, but the result is preceded by the "-" sign. -10 dB means that there was a 10-fold reduction in power compared to the input power.

The relationships apply when comparing currents and voltages

$$dB = 20 \log_{10} \frac{V2}{V1}$$

or

$$dB = 20 \log_{10} \frac{I2}{I1}$$

where

V2 is the output voltage,

I2 is the output current,

and

the base of the logarithm is 20.

It's easy to remember:

Solving problems using electronic circuits

- -6 dB = power gain of 0.251 times (-25.1%)
- -3 dB = power gain of 0.501 times (or roughly 50.1%)
- 0 dB = the output power is equal to the input power. There is no increase or decrease in power.
- 1 dB = power gain of 1.26 times
- 3 dB = power gain of 2 times
- 6 dB = power gain of 4 times
- 10 dB = power gain of 10 times
- 20 dB = power gain of 100 times
- 30 dB = power gain of 1000 times

and

- -6 dB = voltage (current) gain of 0.5 times
- -3 dB = voltage (current) gain of 0.707 times
- 0 dB = voltage (current) gain of 1 time
- 1 dB = voltage (current) gain of 1.122 times
- 3 dB = voltage (current) gain of 1.414 times
- 6 dB = voltage (current) gain of 1.995 times (almost 2 times)
- 10 dB = voltage (current) gain of 3.162 times
- 20 dB = voltage (current) gain of 10 times
- 30 dB = voltage (current) gain of 31.62 times

Addiction C. How to read oscilloscope parameters?

In the book I use a simulator available at https://easyeda.com/. It contains a function generator (symbol on the left). The parameters are visible in brackets above the generator. on the right side you can see the description of the parameters.

Solving problems using electronic circuits

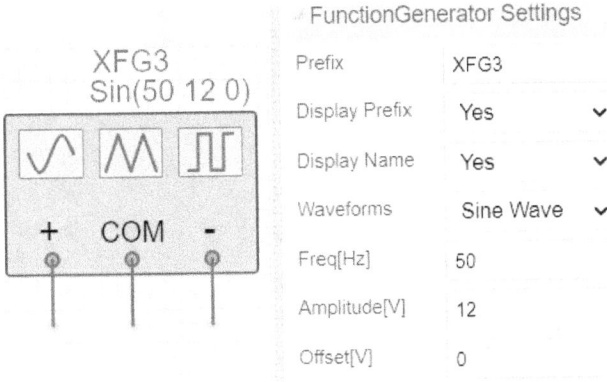

Addition D. Effective value of alternating current

The effective value of alternating current is expressed as the value of direct current that, flowing through a circuit with a constant electrical resistance, would produce the same amount of energy as the given alternating current flowing in the same time. For a sinusoidal waveform, the effective value is related to the peak value by the relationship:

$$U_e = 0.707 U_m$$

$$I_e = 0.707 I_m$$

The effective value of alternating current is denoted in notation by a capital letter with the subscript *e* or — alternatively — by a capital letter without any subscript.

The effective value of the power present in the system is related to the energy lost as heat and defines the requirements that these elements must meet in this respect.

www.ingramcontent.com/pod-product-compliance
Lightning Source LLC
Chambersburg PA
CBHW072054230526
45479CB00010B/1063